Rocks and Soil

written by Maria Gordon
and
illustrated by Mike Gordon

RAINTREE
STECK-VAUGHN
PUBLISHERS
The Steck-Vaughn Company

Austin, Texas

Simple Science

Day and Night
Electricity and Magnetism
Float and Sink
Fun with Color
Fun with Heat
Fun with Light
Fun with Materials
Push and Pull
Rocks and Soil
Skeletons and Movement

Published by Raintree Steck-Vaughn Publishers,
an imprint of Steck-Vaughn Company

Library of Congress Cataloging-in-Publication Data
Gordon, Maria.
Rocks and soil / written by Maria Gordon and illustrated by Mike Gordon.
p. cm.—(Simple science)
Includes bibliographical references (p. -) and index.
Summary: Describes the physical characteristics of rocks and soils and explains how they were formed.
ISBN 0-8172-4504-9
1. Rocks—Juvenile literature.
2. Soils—Juvenile literature.
[1. Rocks. 2. Soils. 3. Rocks—Experiments. 4. Soils—Experiments.
5. Experiments.]
I. Gordon, Mike, ill. II. Title. III. Series: Gordon, Maria. Simple science.
QE432.2.G67 1996
552—dc20 95-41166

Printed in Italy
1 2 3 4 5 6 7 8 9 0 00 99 98 97 96

Contents

Rocks are made of tiny hard pieces, called minerals. They are joined together in lumps.

Soil is a mixture of very small rocks, minerals, pieces of dead plants, dead insects and other animals, water, and air. Most plants grow in soil.

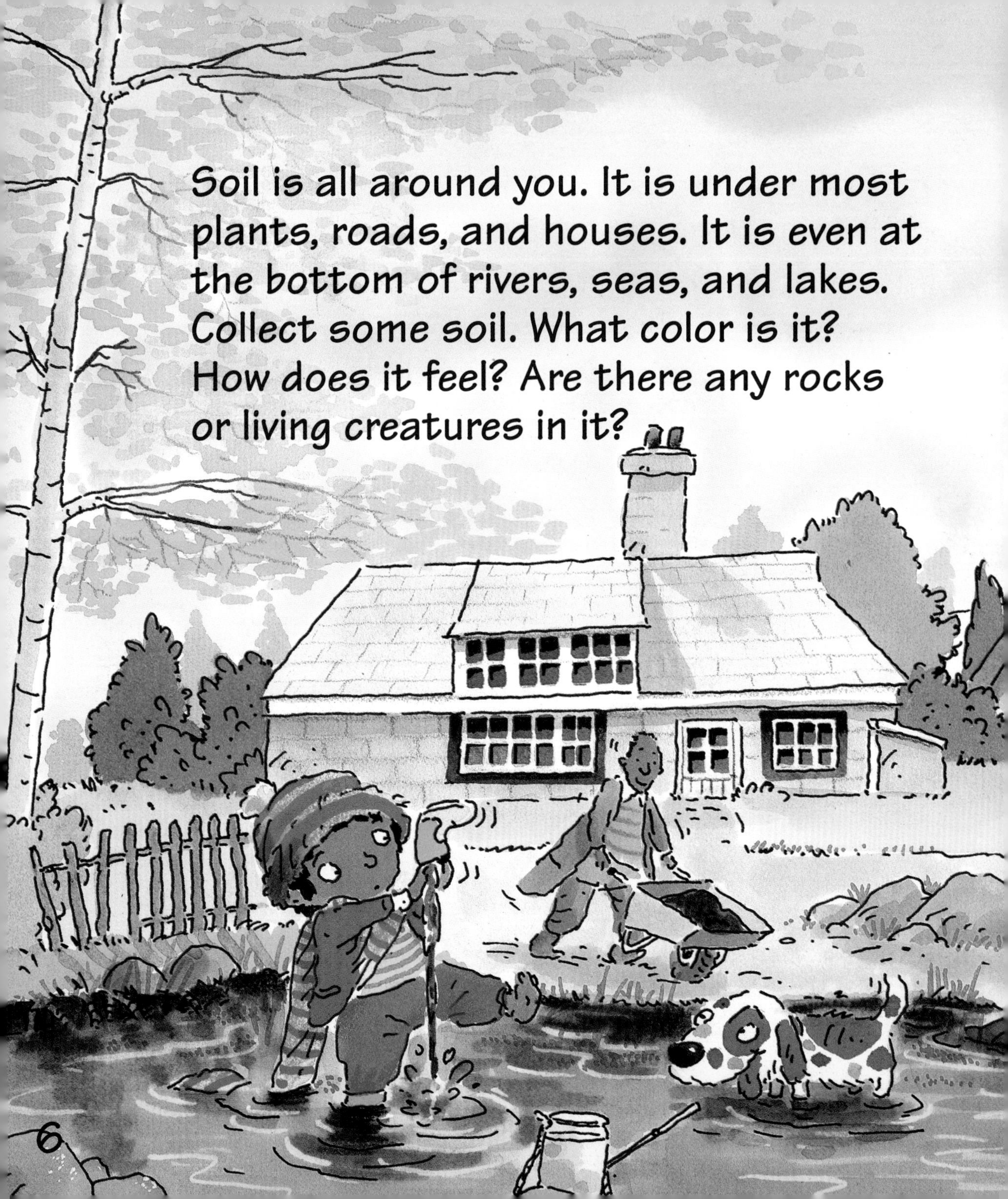

Soil is all around you. It is under most plants, roads, and houses. It is even at the bottom of rivers, seas, and lakes. Collect some soil. What color is it? How does it feel? Are there any rocks or living creatures in it?

You can see rocks in the ground and in cliffs. People make roads, walls, and buildings with rocks. Find some. Are they big or small? How do they feel? What color are they?

Long ago, people used rocks to make tools and weapons. People learned to cook with hot rocks. They made pots out of rocks and built walls and buildings. Beautiful rocks were worn as jewelry.

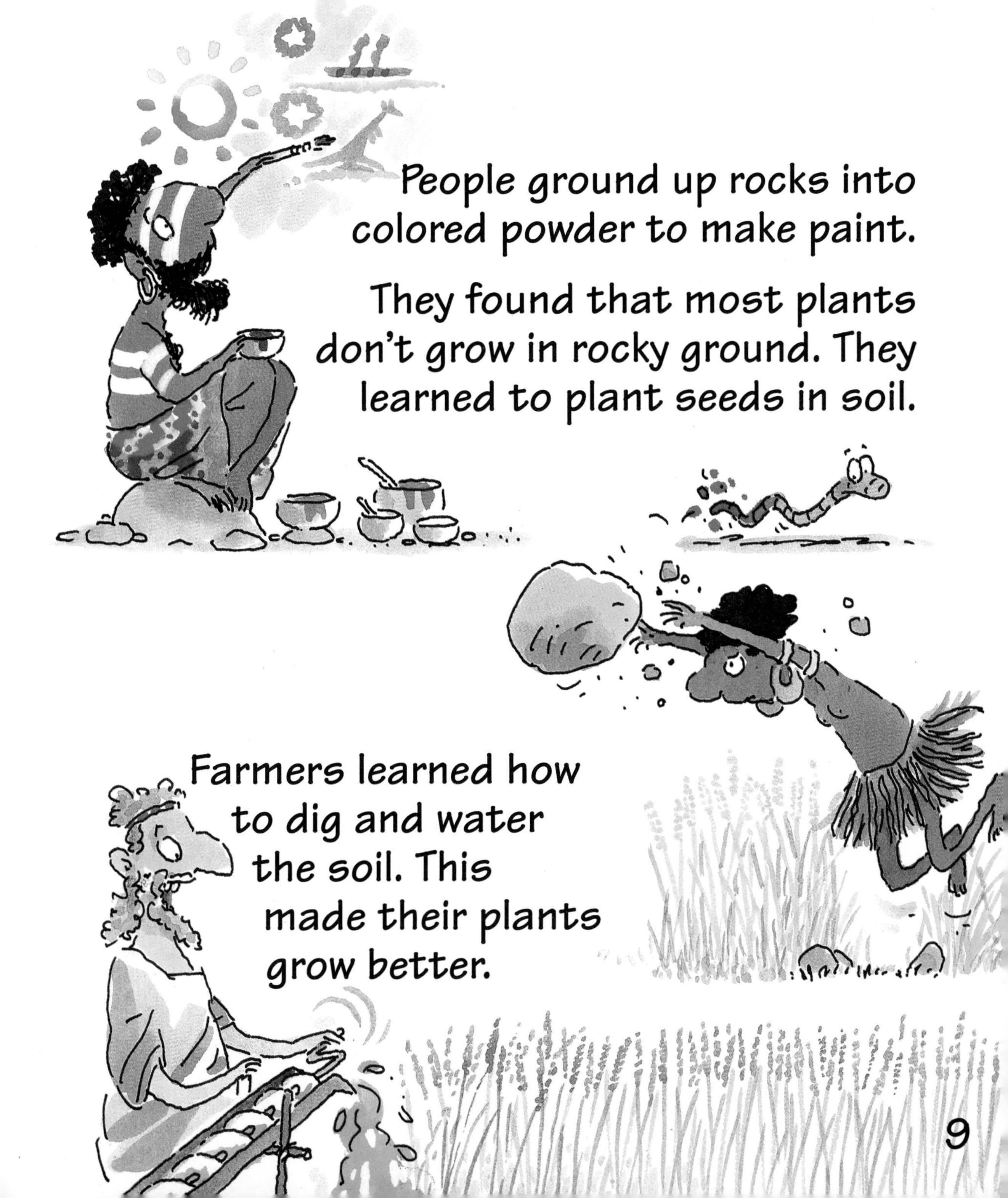

People ground up rocks into colored powder to make paint.

They found that most plants don't grow in rocky ground. They learned to plant seeds in soil.

Farmers learned how to dig and water the soil. This made their plants grow better.

Rocks are made in different ways. Heat makes minerals stick together in lumps. Volcanoes make this kind of rock.

Sugar is like a mineral. Ask an adult to help you melt some—do not touch! It becomes hard when it cools, just like a rock.

Rivers and seas carry minerals like sand and mud to new places. The minerals get dumped on top of each other. This has been happening for millions of years.

Many minerals slowly turn into rock because they are squashed so hard.

Sugar cubes are made like this. They are grains of sugar squashed together.

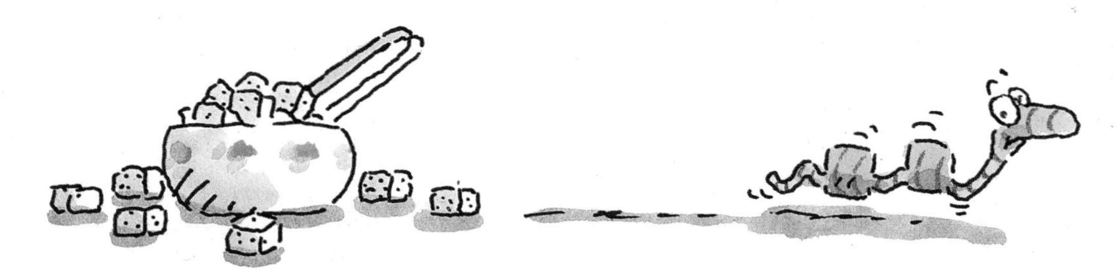

Sometimes minerals fall on dead animals and plants. These animals and plants are slowly pressed onto the sand and mud. This is how fossils are made. Some fossils are the hard parts of animals and plants that were buried. Other fossils are just the shapes the animals made when they were squashed.

Dead leaves, seeds, and branches fall on top of one another. This has happened for thousands of years. Look at some coal. This is made of the plants at the bottom that got squashed and became hard!

Water can carry minerals.
If the water dries up, the
minerals get left behind.

Water dripping in
caves leaves minerals
behind. Very slowly,
this makes big lumps
of rock.

Sometimes you see
mineral deposits in kettles.

Wind, sun, and rain all make rocks break up into little pieces. This is called weathering.

The sun warms rocks. This makes them a tiny bit bigger.

When the rocks cool off at night, they shrink a little. This heating and cooling makes bits of rock crack off. Look for little stones around big rocks.

Water takes up more space when it freezes. Sometimes it freezes inside rocks. The ice pushes against the rocks. This makes them split.

Put some cracked rocks inside a plastic tub.

Cover them with water and freeze them. Then, let the ice melt.

Small pieces of rock may break off into the water. Some cracks will be bigger.

Water rubs away the rough edges of rocks and stones. This makes pebbles and sand. Water carries pebbles and sand, which help to break off big pieces of rock. Waves of water smash rocks, too.

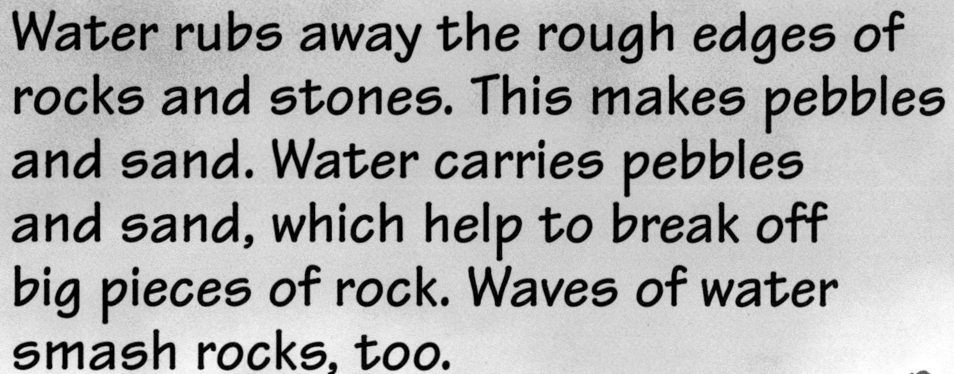

The wind rubs away the softest part of rocks. Plants also break up rocks as they push through any cracks. Even animals can break and crack rocks.

Small pieces of rock help to make soil. Some soils have bigger pieces of rock than others. In sandy soils, you can see little pieces. In silty, or grainy, soils, you need a magnifying glass to see them. In clay soils, the pieces of rock are too small to see.

Next time it rains, see how long it takes the ground to dry. Clay soils become muddy and take a long time to dry.

Sandy soils dry quickly.

Silty soils dry more slowly than sandy soils, but quicker than clay soils.

Ask an adult to help. Find three different soils. Cut out the bottoms of three plastic cups.

Put the cut ends of each cup into the three soils. Pour the same amount of water into each cup. The water soaks fastest into the soil with the most sand in it.

Soil builds up on top of rock. Ask an adult to help you dig down a few inches in a field. See if you hit rock. Watch for worms and insects. Notice where the soil looks darker and where it looks lighter.

The darker part of soil is called topsoil. It has many pieces of plants and animals. The pieces are called humus. They are rotting. This means that creatures in the soil are eating them and turning them into smaller and smaller pieces.

Collect three different kinds of soil in plastic containers.

Put two inches of each type of soil into three different jars.

Fill the jars with water and wait for ten minutes.

Keep the containers of soil—you will need them later!

The biggest pieces of rock sit on the bottom. The tiniest pieces mix with the water and make it cloudy.

Humus floats on top of the water.

Clay soil makes the water cloudiest and silty soil has a lot of humus.

Can you see any little bubbles in the jars? These are filled with air that was in the soil.

Put your tubs of soil in a bright, warm spot. Check them each day for a week. Water them if they become dry and look for seeds sprouting and for tiny living animals.

At the end of the week, plant your own seeds. See which kind of soil they like best.

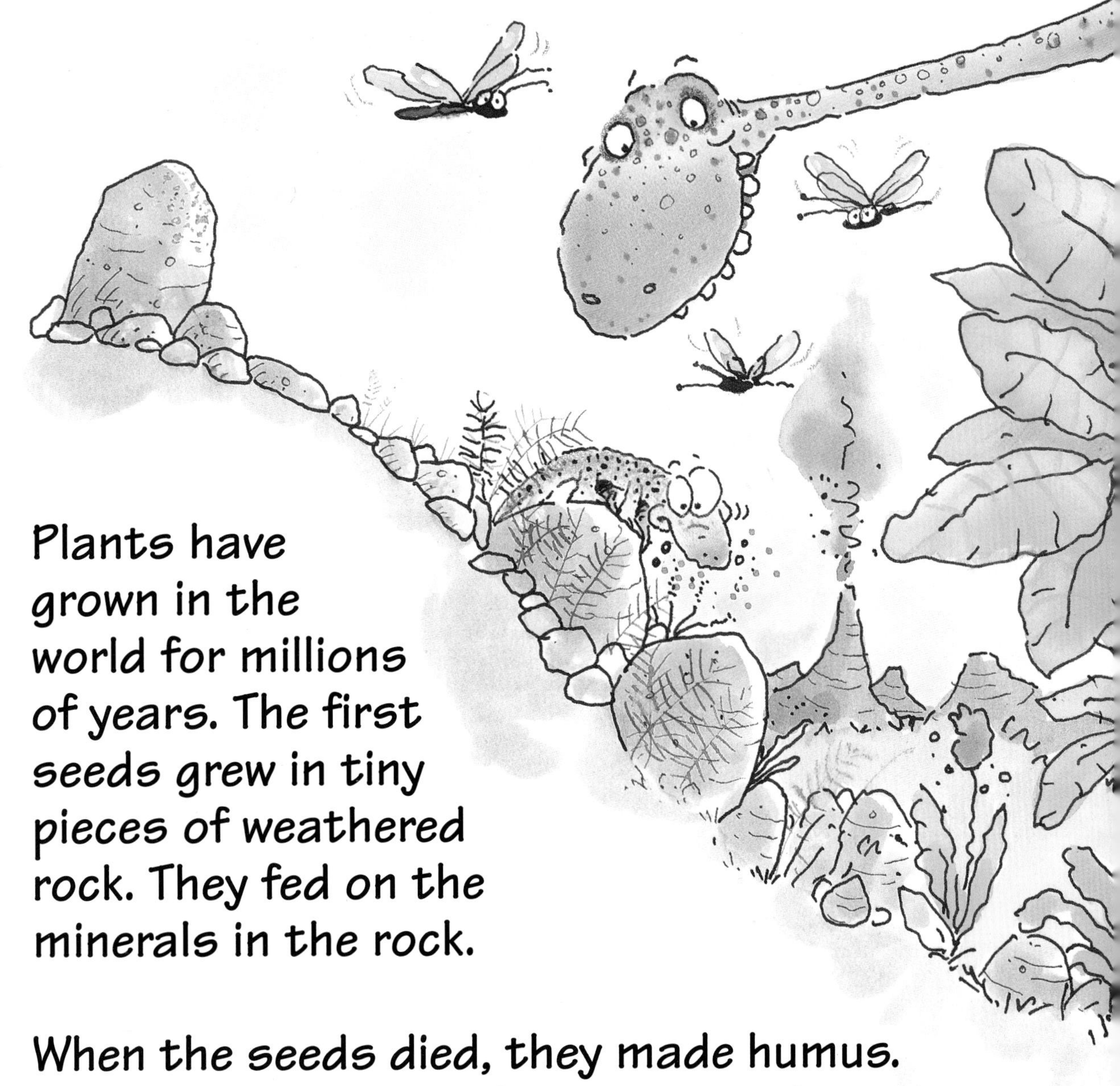

Plants have grown in the world for millions of years. The first seeds grew in tiny pieces of weathered rock. They fed on the minerals in the rock.

When the seeds died, they made humus. The humus was food for tiny animals and more seeds. The mixture of rocks and humus made soil. This took thousands of years.

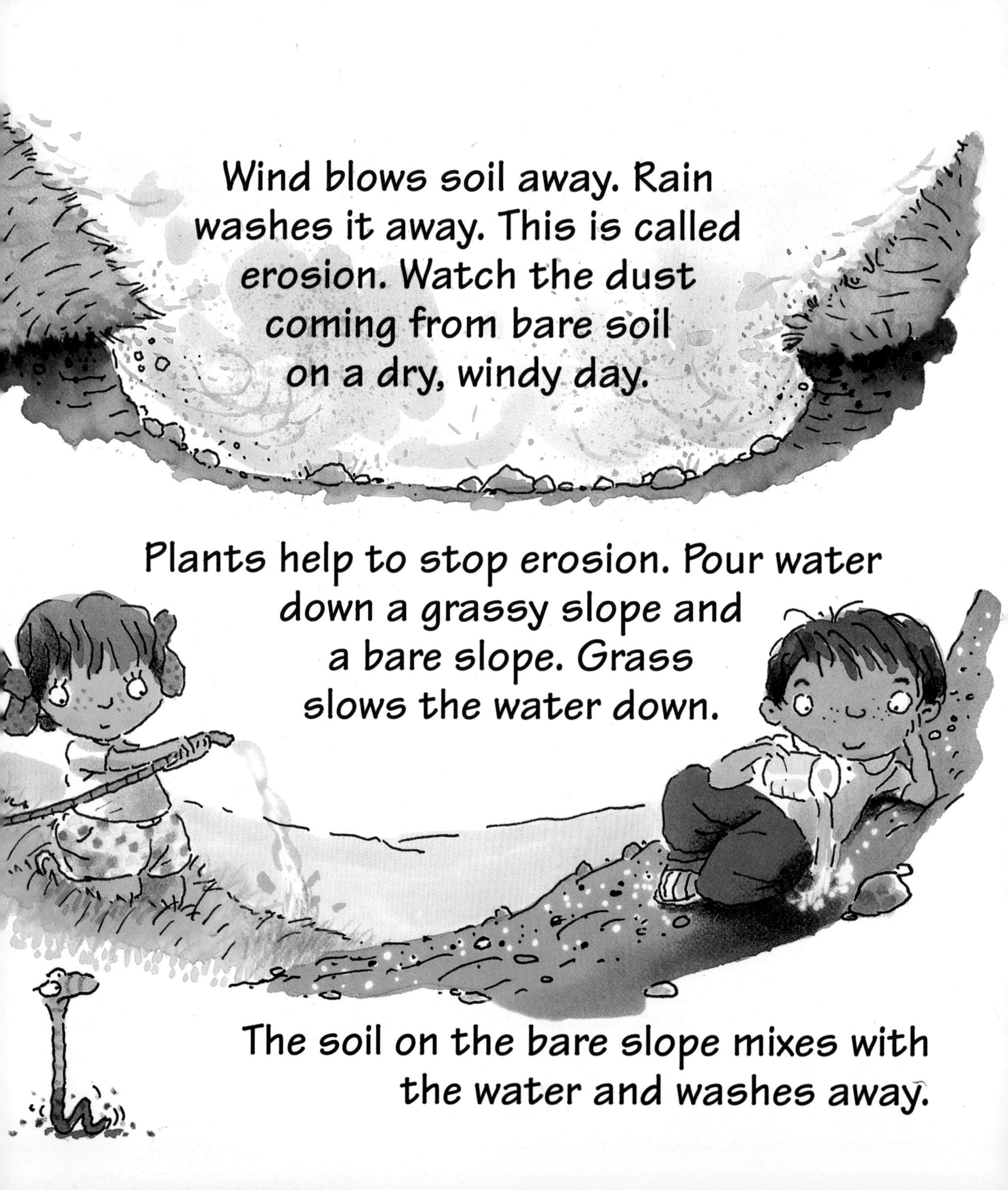

Wind blows soil away. Rain washes it away. This is called erosion. Watch the dust coming from bare soil on a dry, windy day.

Plants help to stop erosion. Pour water down a grassy slope and a bare slope. Grass slows the water down.

The soil on the bare slope mixes with the water and washes away.

Keeping trees, bushes, and plants helps to keep the soil from eroding.

1

Soil needs humus to feed the things we grow. People can save grass clippings, leaves, and old plants to make humus.

Look at the picture. How can you help to make soil and to keep it safe?
The answers are on page 31.

Additional projects

Here are a few more projects to use when examining rocks and soil. The projects go with the pages listed next to them. These projects are harder than the ones in the book, so be sure to ask an adult to help you.

4/5 Make a rock garden. Drag magnets through soil to collect tiny pieces of iron. Find out about the Mohs scale of hardness.

6/7 Visit farms and gardens. Collect and display soil samples.

8/9 Look at sculpture, ancient tools, cave paintings, and jewelry to see how rocks were used a long time ago.

10/11 Use pumice stone (igneous rock). Look at stratification in exposed rock (sedimentary). Grow crystals by dissolving Epsom salts in two glasses of water. Hang a thick string between the glasses, making sure the ends are in both glasses. Put a plate between the glasses, under the string. Leave in place for several days.

12/13 Go fossil hunting.

14/15 Find out how prospectors panned for gold.

16/17 Spot weathering on roads, buildings, etc. Compare shapes of rocks and stones with man-made bricks, etc. Look at pictures of famous gorges and canyons, such as the Grand Canyon.

18/19 Follow paths worn by animals. Find and measure roots breaking through man-made surfaces.

20/21 Make mud pies and sandcastles—compare how much water is needed. Discover which plants grow best in which soils.

22/23 Start or add to a compost heap.

24/25 Investigate soil types of river and pond beds by gently stirring them up.

26/27 Compare mountainside, desert, and river valley vegetation. Look at peat and fertilizers. Mulch some soil. Research crop rotation.

28/29 Examine soil erosion pictures from deforested and dammed sites around the world. See how roots bind soil—for example, on sand dunes.

Answers

1. Start a compost heap. Compost makes fertilizer, which plants need to make food.
2. Plant a tree. Trees prevent erosion by binding soil and by blocking wind.
3. When a sign tells you to stay off the grass, stay off the grass! It may be that the grass is newly planted and needs a chance to grow.

Other books to read

Barkan, Joanne. **Rocks, Rocks Big and Small.** First Facts. New York: Silver Press, 1990.

Butler, Daphne. **First Look Under the Ground.** First Look. Milwaukee: Gareth Stevens, 1991.

Fowler, Alan. **It Could Still Be a Rock.** Rookie Read-About Science. Chicago: Childrens Press, 1993.

Lye, Keith. **Rocks and Minerals.** First Starts. Milwaukee: Raintree Steck-Vaughn, 1992.

Stille, Darlene. **Soil Erosion and Pollution.** New True Books. Chicago: Childrens Press, 1990.

Index